Waves

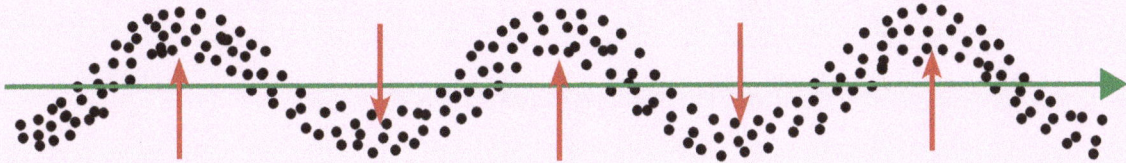

Illustrations: Janet Moneymaker
Design/Editing: Marjie Bassler

Waves
ISBN 978-1-950415-26-7

Published by Gravitas Publications Inc.
Imprint: Real Science-4-Kids
www.gravitaspublications.com
www.realscience4kids.com

What happens when you put your foot in the bathtub?

You can see a ripple of water move away from your foot.
This ripple is called a **wave.**

Wavy!

A wave happens when **energy** is transferred by **vibration.** A **vibration** occurs when something moves back and forth.

When you put your foot in water, the water molecules surrounding your foot begin to move back and forth, or **vibrate.** The vibrating water molecules transfer energy to other water molecules and make them vibrate too. This creates a wave.

Wavy!

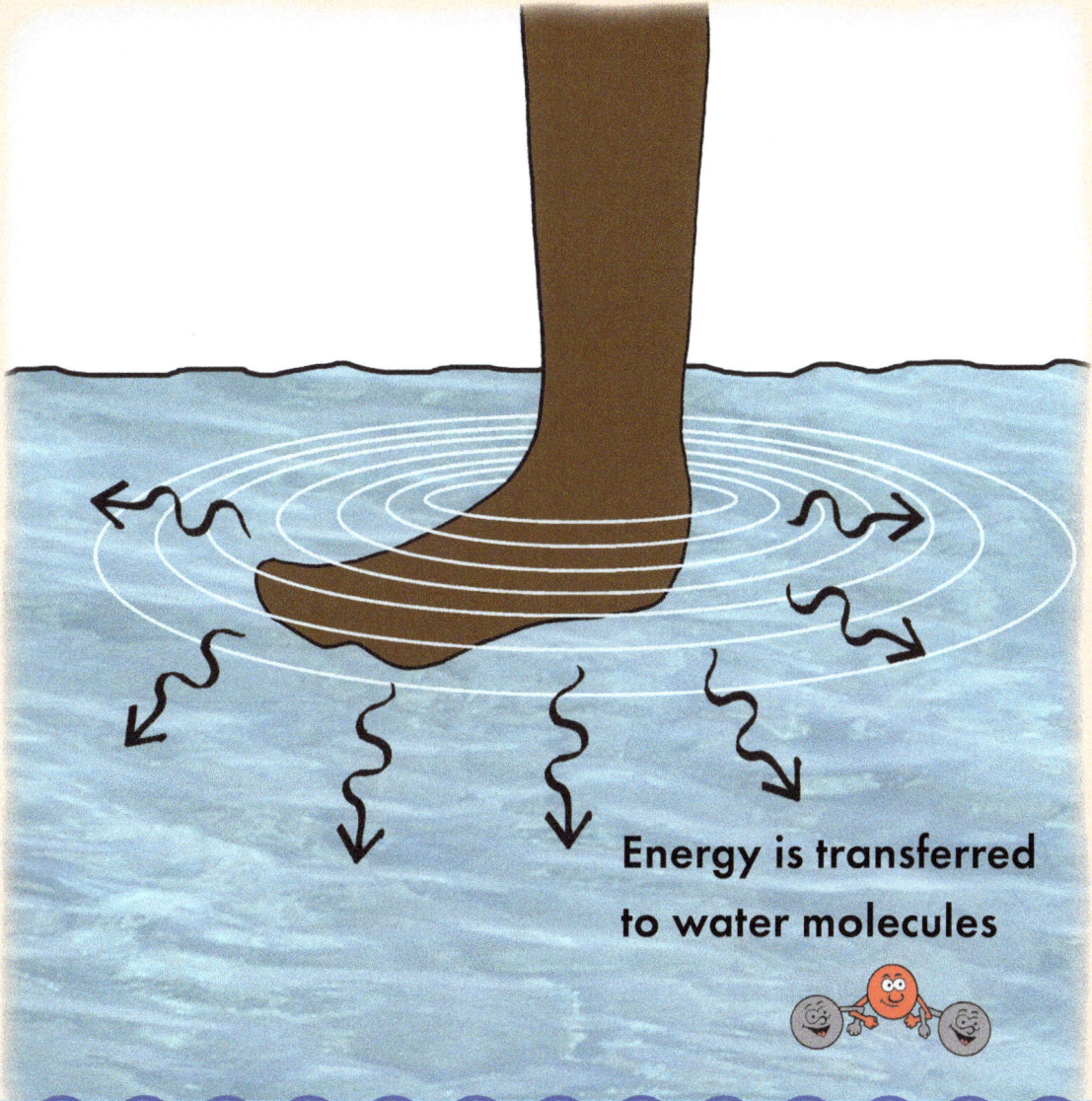

Energy is transferred
to water molecules

Energy is needed to do **work**.

Work happens when a **force** moves an object.

Force is any action that changes...

...the **location** of an object,

...the **shape** of an object,

...**how fast or how slowly** an object is moving. (This is called the **speed** of an object.)

Atoms are tiny building blocks that can link together.

Atoms make up everything we touch, taste, smell, and see.

Molecules are made when **atoms link** together.

All waves have a particular shape.

Waves have a high point called a **peak**.

Waves also have a low point called a **valley**.

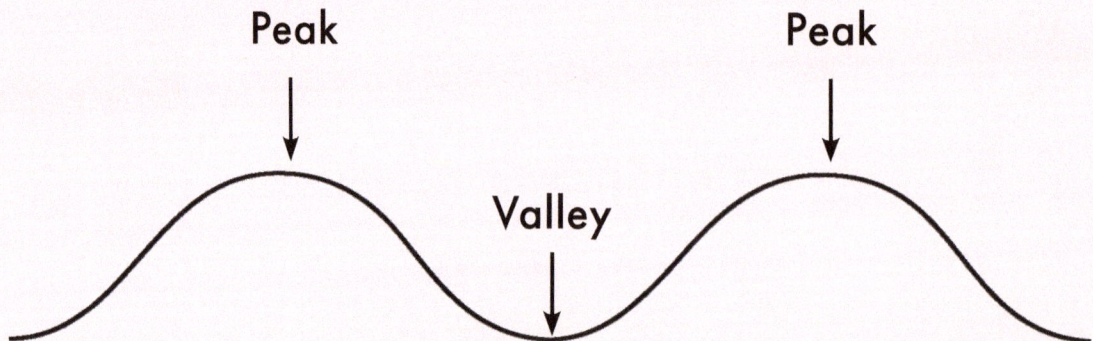

Peak

Peak

Valley

The size of a wave depends on how much energy it has.

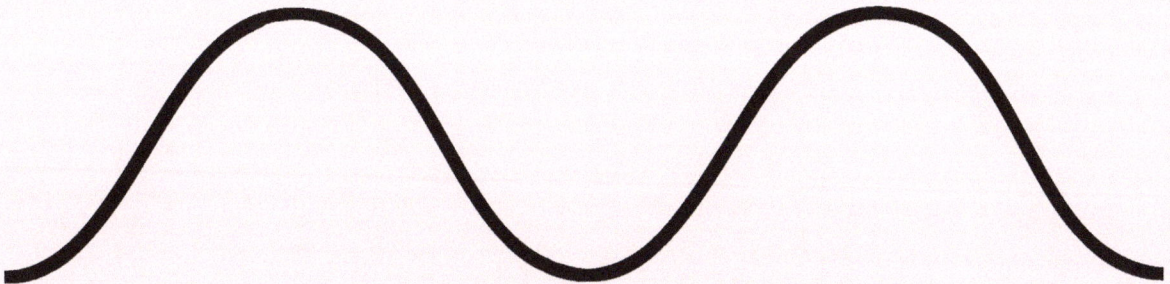

Big Wave = More Energy

Small Wave = Less Energy

You might think the water in the wave is moving forward, but IT IS NOT!

The water molecules actually stay in one place.

Molecules simply move up and down or slightly side to side during a wave.

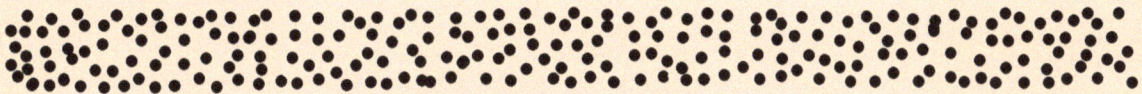

Molecules before a wave passes through them.

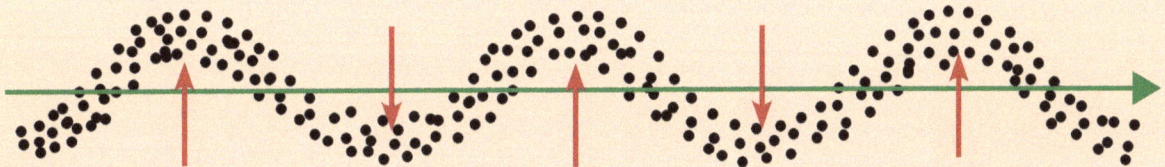

Molecules move up and down as a wave passes through them.

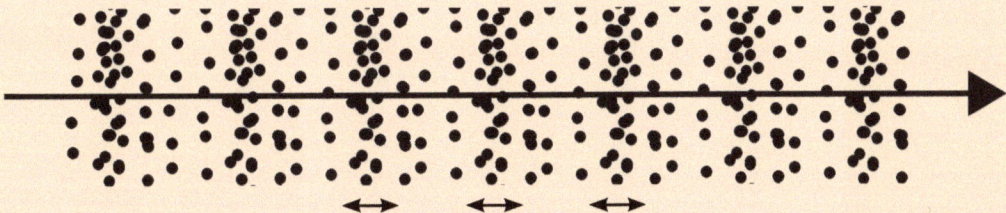

Molecules move side to side as a wave passes through them.

An up and down wave is
called a **transverse wave.**

Transverse Wave

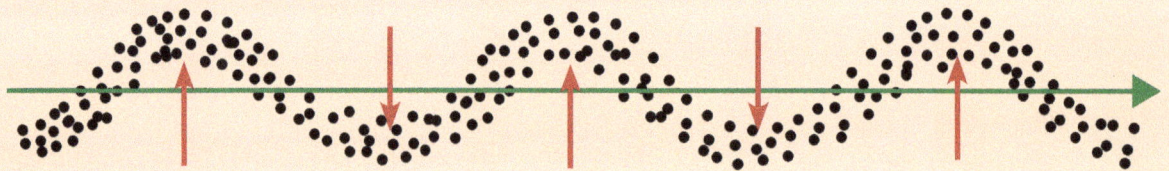

Molecules move up and down as a wave passes through them.

A good example of a transverse wave is a stadium wave.

The stadium wave moves through the crowd because the people stand up and sit down. But they do not move from one chair to another.

In a similar way, the molecules in a transverse wave move up and down, but they do not change position.

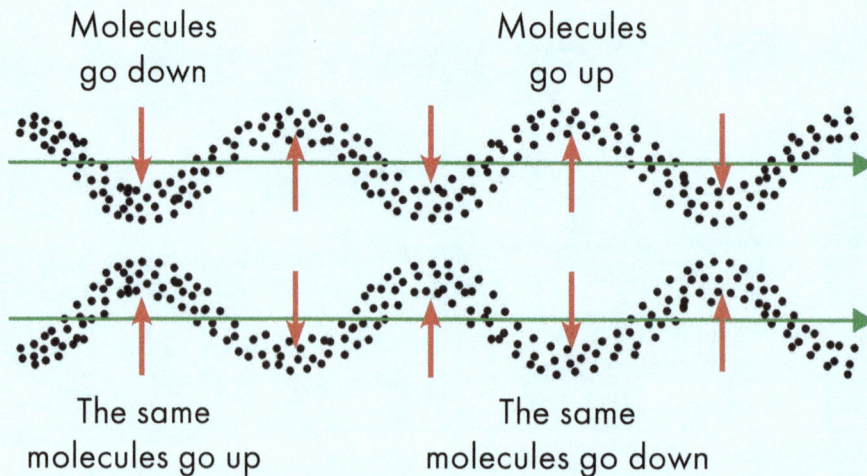

Molecules go down

Molecules go up

The same molecules go up

The same molecules go down

A wave where the molecules move slightly side to side is called a **longitudinal wave**.

A good example of a longitudinal wave is the movement of a spring. When you pull the spring and let it go, the energy will travel down the coils of the spring making them move side to side.

Longitudinal Wave

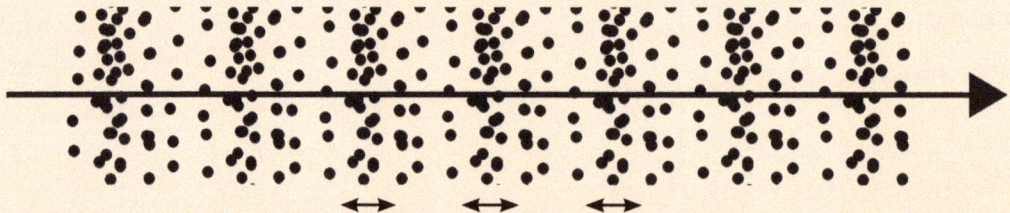

Molecules move side to side as a wave passes through them.

If you push part of the spring together and then let go, energy will also travel along the spring to make the coils move side to side.

A wave moves through material by changing the energy of the molecules!

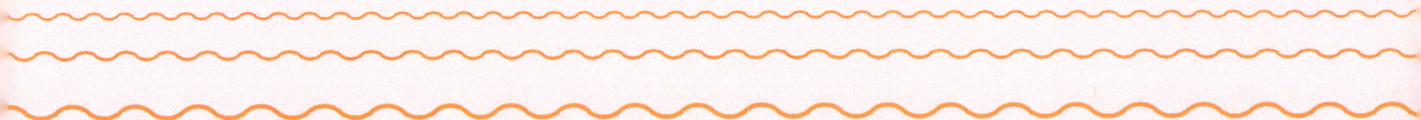

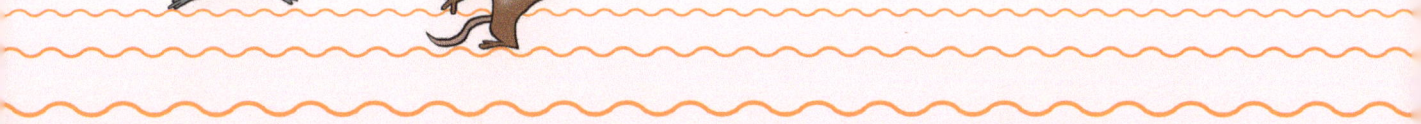

How to say science words

atom (AA-tum)

energy (E-nuhr-jee)

force (FAWRSS)

location (loh-KAY-shun)

longitudinal (lawn-juh-TOOD-nuhl)

molecule (MAH-lih-kyool)

peak (PEEK)

shape (SHAYP)

speed (SPEED)

transverse (tranz-VERSS)

valley (VAA-lee)

wave (WAYV)

work (WERK)

What questions do you have about WAVES?

Learn More Real Science!

Complete science curricula from Real Science-4-Kids

Focus On Series

Unit study for elementary and middle school levels

Chemistry
Biology
Physics
Geology
Astronomy

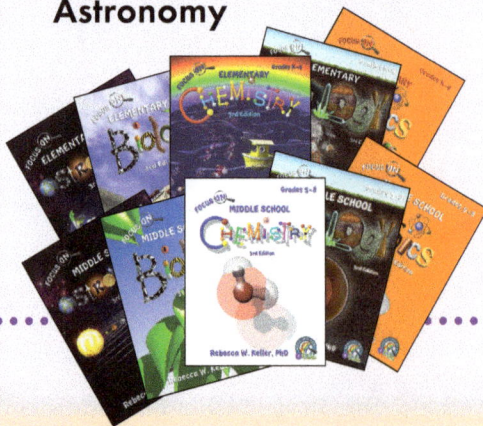

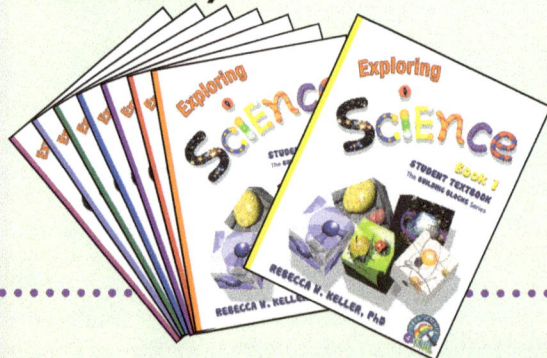

Exploring Science Series

Graded series for levels K–8. Each book contains 4 chapters of:

Chemistry
Biology
Physics
Geology
Astronomy

www.ingramcontent.com/pod-product-compliance
Lightning Source LLC
Chambersburg PA
CBHW040153200326
41520CB00028B/7590